Impacts of Climate Change and Variability on Transportation Systems and Infrastructure

The Gulf Coast Study, Phase 2

Assessing Transportation System Vulnerability to Climate Change: Synthesis of Lessons Learned and Methods Applied

Final Report, Task 6

Prepared for:

The USDOT Center for Climate Change and Environmental Forecasting

Project managed by:

Office of Planning, Environment, and Realty
Federal Highway Administration, USDOT

Prepared by:

ICF International
1725 Eye Street, NW
Washington, DC 20006

Date: **October 2014**

Contract No.: **GS-10F-0124J**   Report No.: **FHWA-HEP-15-007**

Acknowledgements

Principal Authors and Reviewers (in alphabetical order):

Ms. Anne Choate, Ms. Brenda Dix, Ms. Jessica Kuna, Ms. Beth Rodehorst, Ms. Cassandra Snow, of ICF International; Brian Beucler, Robert Hyman, Robert Kafalenos, of Federal Highway Administration (FHWA)

Contents

Figures

Tables

1. Project Background

The U.S. Department of Transportation (U.S. DOT) conducted a comprehensive, multi-phase study of the Central Gulf Coast region to better understand climate change impacts on transportation infrastructure and identify potential adaptation strategies. This region is home to a complex multimodal network of transportation infrastructure, and it plays a critical economic role in the import and export of oil and gas, agricultural products, and other goods. Phase 1 of this Gulf Coast Study (completed in 2008) examined the impacts of climate change on transportation infrastructure at a regional scale.

> **Project Objectives**
>
> - Develop and test methodologies for evaluating the vulnerability of a metropolitan transportation system to climate change
>
> - Use lessons learned in Mobile to develop transferrable screening tools and approaches to help other regions identify vulnerabilities and consider options for protecting/adapting

Phase 2 (completed in 2014) focused on a smaller region, enhancing regional decision makers' ability to understand potential impacts on specific components of infrastructure and to evaluate adaptation options. An important goal of Phase 2 was to develop methodologies that could be used by other transportation agencies to evaluate vulnerability and adaption measures. With that goal in mind, the project team developed transferrable methodologies and pilot tested them on the transportation system in Mobile, Alabama.

This study evaluated the impacts on six transportation modes (highways, ports, airports, rail, transit, and pipelines) from projected changes in temperature and precipitation, sea level rise, and the storm surges and winds associated with more intense storms. The project resulted in findings on Mobile's transportation vulnerability, as well as approaches for using climate data in transportation vulnerability assessments, methods for evaluating vulnerability and adaptation options, and tools and resources that will assist other agencies in conducting similar work.

This report provides a synthesis of the Phase 2 study, including:[1]

- Task 1: Evaluate Criticality

- Task 2: Gather and Process Climate Information

- Task 3: Assess Vulnerability

---

[1] Full reports are available for tasks 1 through 3 on FHWA's Gulf Coast Study web site at
http://www.fhwa.dot.gov/environment/climate_change/adaptation/publications_and_tools/vulnerability_assessment_framework/page01.cfm.
These reports contain detailed methodology and results. The tools and resources developed under Task 4 are housed on FHWA's virtual Climate Change and Extreme Weather Vulnerability Assessment Framework at
http://www.fhwa.dot.gov/environment/climate_change/adaptation/adaptation_framework. No formal reports or tools were developed under Task 5.

- Task 4: Develop Tools and Resources

- Task 5: Coordinate with Planning Authorities and the Public

This report constitutes Task 6. It discusses highlights of the project and summarizes the methodologies employed and resulting tools. Moreover, it discusses how these methodologies may be applied by other transportation agencies wishing to conduct similar vulnerability assessments, providing information on resources available to assist other agencies at each stage of a vulnerability assessment. Finally, this report discusses areas for future work.

Throughout the report, Mobile-specific findings are described in shaded boxes; these examples illustrate how higher-level findings relate to the specific assets and services studied in Mobile.

2. Project Highlights

Phase 2 of the Gulf Coast Study is meant to help transportation agencies overcome common resource and knowledge constraints that make it difficult to incorporate climate change into their planning, design, and operations and maintenance activities. The project yielded important lessons learned, methodologies, and tools that may assist other transportation agencies in conducting similar vulnerability assessments. Some project highlights include:

- **Lessons learned about developing and using detailed, downscaled climate projection information.** Detailed, local climate projection information is an important tool when evaluating how climate change may affect transportation. However, there are different ways to develop this information, and countless formats that the resulting data could take (e.g. presenting temperature projections in terms of changes in seasonal averages, or in terms of the number of days above 95 degrees). This project developed lessons learned regarding how to develop projections that are relevant to transportation, and how to harness the range of plausible projections into more manageable "climate narratives" against which to evaluate vulnerability and inform asset-level analyses.

- **Methodologies to screen transportation assets for criticality and vulnerability.** The methodologies developed for this project allow for effective and efficient assessments of a large number of transportation assets within a system. These methodologies can be adjusted to account for the range of information available, so that any transportation agency could use them to screen their transportation assets for criticality and vulnerability.

- **Approaches for conducting detailed engineering analyses on individual assets for a range of modes and climate stressors, to better understand their specific vulnerabilities and options for adaptation.** This project included project-specific vulnerability and adaptation analyses. Individual assets were assessed for their vulnerability to particular climate change stressors, and feasible adaptation options were evaluated. The methodologies developed for these analyses could be adapted and employed by other transportation agencies wishing to evaluate vulnerabilities and adaptation measures for specific transportation projects.

- **Tools to assist other transportation agencies in conducting similar assessments**. The tools developed for this study include an Excel-based Vulnerability Assessment Scoring Tool (VAST) to simplify vulnerability assessments; a Coupled Model Intercomparison Project (CMIP) Climate Data Processing Tool to "translate" projected changes in local temperature and precipitation into terms that are relevant to transportation stakeholders; a Sensitivity Matrix that shows key sensitivities of transportation assets to each climate stressor; a guidance document for *Assessing Criticality in Transportation Planning;* and a web-based framework for evaluating vulnerability, with various videos, reports, and other resources to assist transportation practitioners at each stage of their assessments.

3. Overview of Methodology and Final Tools

This project screened and analyzed transportation assets in Mobile to evaluate system-level and asset-level vulnerabilities to climate change. First, a scoping step was conducted to determine project boundaries. The assets were then screened for criticality, and then screened and analyzed for vulnerability. Finally, a small subset underwent detailed engineering analyses.

The scoping process focused the project within certain boundaries, including:

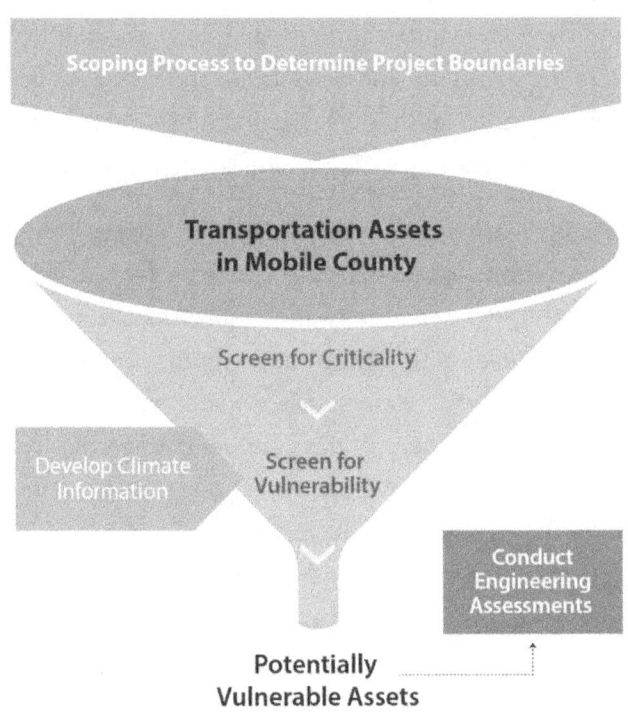

Figure 1: Graphical representation of the vulnerability screen process employed in this study

- The geographic boundary (Mobile County plus the bridge touch-downs on the other side of the Mobile Bay; offshore facilities were excluded),

- Transportation modes to evaluate (highways, ports, airports, rail, transit, pipelines),

- Types of asset within each mode to evaluate (e.g. roadway classifications contained in the regional long-range transportation plan, not local roads),

- Climate change stressors, scenarios, and timeframes to consider.

This scoping process was important, and has notable implications on the results. For example, there are a number of potential transportation disruptions that occur outside the boundaries of Mobile County, but that could affect the Mobile community (e.g. damage to rail lines outside of Mobile that prevent trains from using the rail lines that run through Mobile). Furthermore, the disruption of Mobile's transportation system could have major repercussions in other parts of the country; e.g., if goods cannot be shipped to/from the Port of Mobile. This study looked only at impacts on *Mobile's* transportation system, and how the *Mobile* community would be affected by those impacts. However, the geographic boundaries could have been drawn to look at impacts in the larger region from disruptions to Mobile's transportation system. In this situation, different assets might have appeared to be more critical or more vulnerable.

The approach was also designed to whittle down the large number of transportation assets in Mobile, Alabama to a subset of assets that are both highly critical and vulnerable to extreme weather and climate change (see Figure 1). The goals of this screening approach were twofold: identify system-level insights into overarching vulnerabilities in Mobile, and identify specific assets that appear to be vulnerable to specific climate stressors. A subset of assets then underwent a detailed engineering assessment to investigate possible vulnerabilities, and to explore adaptation measures. Finally, a series of tools were developed—based on the methodologies used in this project—to assist other transportation agencies in conducting similar assessments.

A screening approach was taken because it is not feasible to conduct detailed, project-level vulnerability assessments on the hundreds or thousands of assets that comprise a transportation system; this is true in Mobile, AL, New York, NY, and for almost all other transportation systems. Using high-level screening techniques, transportation practitioners can hone in on the assets that are most critical to their communities, and most likely to be vulnerable to climate change, before taking a detailed look at a small number of assets. The screen is also an effective tool for identifying system-wide vulnerabilities.

An overview of each step of the screening process is discussed below. For more detail on the methodologies, please see the technical reports on FHWA's Gulf Coast Study web site.[2]

3.1. Screen for Criticality

The approach used in the Gulf Coast project first screened the assets within Mobile's transportation system for criticality. Criticality refers to an asset's importance to the local transportation system, economy, and general functioning of the community.[3] To conduct this criticality assessment, the project team developed a scoring system that ranked each asset's criticality as High, Medium, or Low. To do so, a set of criteria for evaluating criticality was developed. The specific criteria varied for each mode, but all criteria related to *socioeconomic* importance, *use and operational* characteristics, or the *health and safety* role in the community. These criteria were scored using methods ranging from statistics on use (such as volume of cargo throughput at a port), to traffic modeling to determine the level of redundancy and the impact on the system if a particular segment were to become inaccessible, to expert

---

[2] Available at http://www.fhwa.dot.gov/environment/climate_change/adaptation/ongoing_and_current_research/gulf_coast_study/

[3] Criticality is not a measurement of vulnerability. An asset can be highly critical to the local community, but not vulnerable to climate change. Conversely, an asset could be vulnerable to climate change, but not critical to the community. The latter type of asset would not be evaluated under the methodology discussed in this report, since this asset would be screened out under the criticality assessment and therefore not evaluated for vulnerability.

judgment. The process is meant to be scalable to the system being examined. The scores were then averaged to determine an overall criticality score. The project team used these scores to select the most critical assets across different modes to evaluate for vulnerability.

3.2. Gather and Process Climate Information

It is difficult to assess vulnerability without an understanding of how the climate may change in the future. Therefore, the project developed climate information relevant to transportation planners to characterize plausible future climate scenarios in Mobile. Table 1 on page 16 summarizes the climate stressors, scenarios, and timeframes used for projecting future climate conditions in Mobile.

3.3. Screen for Vulnerability

Several hundred assets or segments were considered to be "highly critical," and detailed vulnerability assessments could not be conducted on each asset. Therefore, this study identified appropriate "indicators" of the three components of vulnerability:[4]

- **Exposure**—Extent to which an asset experiences climate variability and change

- **Sensitivity**—Degree to which an asset is affected by exposure (i.e., if all assets were equally exposed, which would experience the greatest damage?)

- **Adaptive capacity**—Ability of a system to adjust, repair, and flexibly respond to damage to an asset.

An indicator is a representative data element that can be used as a proxy measurement of the overall exposure, sensitivity, and adaptive capacity of specific assets. For example, paving materials vary in their sensitivity to temperature, so looking at the types of paving materials used for highways or runways can provide an indication of how sensitive specific assets might be to high temperatures.

Indicators were scored on a scale of 1 through 4, and then a composite vulnerability score was calculated for each asset. Assets received a vulnerability score for each of the five climate stressors studied.

An essential step of the vulnerability assessment was the review of initial results by local stakeholders, who could put the results in the context of their knowledge of the transportation assets and the Mobile area. These discussions led to adjustments in the list of indicators, the scoring of indicators, assumptions about thresholds of sensitivity, and discussions about the

---

[4] Definitions for vulnerability, exposure, sensitivity, and adaptive capacity are adapted from the International Panel on Climate Change (IPCC)'s Third Assessment Report. Please see "Annex B: Glossary of Terms" at http://www.ipcc.ch/pdf/glossary/tar-ipcc-terms-en.pdf.

regional transportation system's ability to "bounce back" after a particular asset fails. These refinements yielded more robust results, strengthening the quality of the assessment.

3.4. Conduct Detailed Engineering Assessments

The project team then took a closer look at a small number of the transportation assets through a series of engineering case studies.[5] Zeroing in on a specific feature of the asset (e.g., the embankment of a roadway) and a particular climate stressor (e.g., storm surge), these detailed analyses considered the engineering design specifications and potential failure pathways, and then evaluated how the asset might be vulnerable to the climate stressor. Evaluations of specific potential adaptation options (i.e., options for reducing vulnerability) were also conducted. This work represents some of the most detailed assessments to date of transportation vulnerability and adaptation for a wide range of transportation assets. Each of these analyses (11 in all) comprises an individual case study based on unique methodologies and results.

---

[5] For purposes of this study, asset-stressor combinations were selected to represent a wide variety of challenges facing engineers when designing for future climate impacts; not all of the assets were considered to be highly vulnerable according to the vulnerability screen. When conducting their own vulnerability assessments, however, transportation practitioners would likely want to conduct detailed assessments on assets that score high in the vulnerability screen.

4. Tools and Resources

The Gulf Coast study developed tools and resources to help other agencies capitalize on the methods developed and tested under this project. This section summarizes those resources.

4.1. The Virtual Framework

The Virtual Framework is an interactive web version of the *Federal Highway Administration* (*FHWA*) *Climate Change and Extreme Weather Vulnerability Assessment Framework* (Figure 2). The Virtual Framework breaks the vulnerability assessment process into six modules. Each module contains step-by-step guidance, video testimonials from professionals sharing lessons on their experience, case studies related to the framework step, links to resources related to the step, and tools to help a user complete the step.

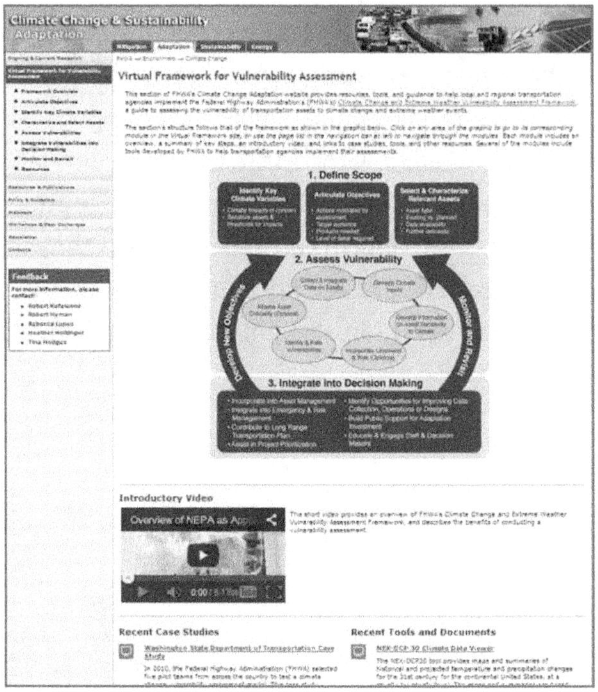

Figure 2: Screen shot of FHWA's web-based Virtual Framework

4.2. Guidance and Tools

The methodology and lessons learned from the Gulf Coast Study were the basis for several tools and resources that will assist other transportation agencies in conducting vulnerability assessments and the steps of FHWA's Virtual Framework.

- **Assessing Criticality in Transportation Adaptation Planning** provides a step-by-step guide to assessing criticality, using examples from both the Gulf Coast Study and other projects. This guide provides tips on scoping and defining criticality and applying criteria and ranking assets.

- **Transportation Climate Change Sensitivity Matrix** is an Excel tool that helps users identify how different climate stressors (ranging from drought to storm surge) affect various types of transportation infrastructure.

- **CMIP Climate Data Processing Tool** is an Excel tool that processes downscaled climate data at the local level into dozens of specific temperature and precipitation variables relevant to transportation agencies. With this tool, users can get projections for these specific, transportation-relevant climate variables at the local level in a matter of hours. For examples of the variables included in this tool, please see the textbox on page 20. Figure 3 shows an example input tab.

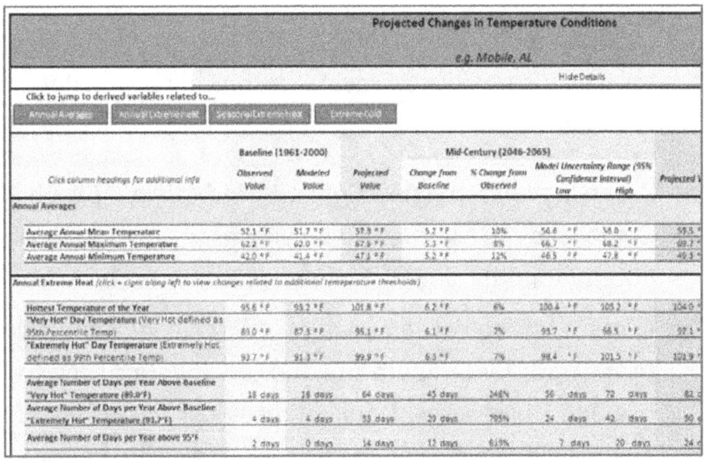

Figure 3: Example Input Sheet from CMIP Climate Data Processing Tool

The tool relies on statistically downscaled climate data from the U.S. Bureau of Reclamation's Downscaled CMIP3 and CMIP5 Climate and Hydrology Projections website. The data are available at the 1/8 degree resolution (roughly 56 sq miles for the US) covering the contiguous United States. [6]

- **Vulnerability Assessment Scoring Tool (VAST)** is an Excel spreadsheet that walks users through a step-by-step process to conduct an indicator-based vulnerability screen (Figure 4). VAST takes a user through the key steps and questions necessary to conduct a vulnerability assessment, such as choosing the climate stressors and assets to evaluate, selecting indicators, collecting data on those indicators, developing an approach to convert indicator data into vulnerability scores,

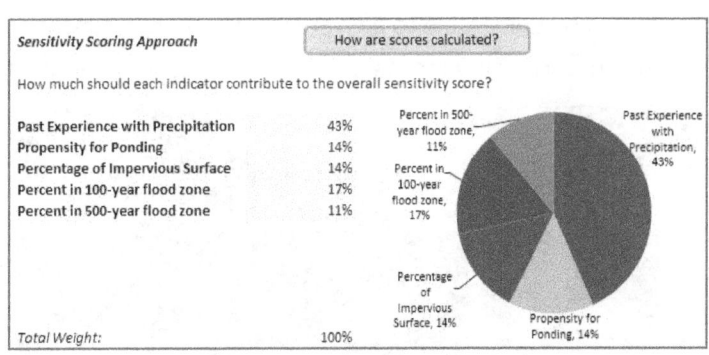

Figure 4: Example Input Sheet from VAST

and reviewing the results. The tool is designed to be flexible, and it allows users to use any indicators, data sources, and climate stressors they choose. VAST includes hundreds of example indicators and example scoring approaches that can be utilized.

---

[6] See http://gdo-dcp.ucllnl.org/downscaled_cmip_projections/dcpInterface.html

4.3. Technical Reports

The project also produced several technical reports documenting all methodologies tested
throughout the study:

a. *Assessing Infrastructure Criticality in Mobile, AL* details the methodology used in the Gulf
   Coast Study to conduct a criticality assessment of Mobile's transportation assets. Example
   criteria for criticality of each mode are included in the report.

b. *Climate Variability and Change in Mobile, AL* details the methodology used to develop
   projections for temperature and precipitation, and to model the inundation from sea level
   rise and storm surge. Detailed results of the projection and modeling efforts are included.[7]
   The methodology used in this report laid the groundwork for the CMIP Climate Data
   Processing Tool.

c. *Assessing the Sensitivity of Transportation Assets to Climate Change* and accompanying
   Sensitivity Matrix discuss how different transportation modes and asset types in Mobile are
   sensitive to climate stressors. The *Transportation Climate Change Sensitivity Matrix* tool was
   developed based on this report and accompanying Matrix.

d. *Screening for Vulnerability* covers the detailed methodology used to conduct the
   vulnerability screen. Example indicators, data sources, and scoring approaches are included.
   The methodology used in this report is the foundation of VAST.

e. *Engineering Assessments of Climate Change Impacts and Adaptation Measures* presents 11
   engineering case studies that took a detailed look at specific assets in Mobile and how they
   would be vulnerable to a given climate change stressor, as well as potential adaptation options.
   This report includes detailed methodology for conducting the engineering assessments.

---

[7] Information on the modeling approach and results is also available from "U.S. DOT. 2012. Temperature and Precipitation Projections for the
Mobile Bay Region. The Gulf Coast Study, Phase 2, Impacts of Climate Change and Variability on Transportation Systems and Infrastructure.
Prepared by Katharine Hayhoe and Anne Stoner of Texas Tech University Climate Science Center for the U.S. DOT Center for Climate Change and
Environmental Forecasting, FHWA-HEP-12-055." Available at:
http://www.fhwa.dot.gov/environment/climate_change/adaptation/ongoing_and_current_research/gulf_coast_study/phase2_task2/mobile_infrastructure.

5. Applying the Gulf Coast Study Methodologies and Lessons Learned

5.1. Evaluate Criticality

5.1.1. Overview of Approach

As the Mobile transportation system contains thousands of individual assets (i.e., road and rail segments, ports, airports, etc.) , it was useful to first identify which assets were considered highly critical to the community, and then focus the vulnerability assessment on those assets.

What does "critical" mean? There is no single definition. In the context of vulnerability assessments, "critical" should reflect the priorities of the community or group conducting the study. For example, some stakeholders may value connections to areas that are vital to the local economy or identity, while other stakeholders may place more value on the role that transportation assets play in state or regional connections. The Gulf Coast Study focused on assets critical to Mobile; however, studies could also focus on state or national priorities (e.g., national freight flows). Ultimately, criteria can be adjusted to reflect study goals.

Thus, a first step when evaluating criticality is to define the qualities that might make transportation assets highly critical for the community. **Defining criticality is an exercise that depends on many factors, including local priorities, data availability, and even the definition of each "asset".** There is no prescriptive way to define criticality, and it is important to spend time upfront discussing the intended scope and goals of the effort. The appropriate evaluation criteria for ranking criticality will vary depending on the intended audience and goals of the study, what resources are available, and other factors.

For the Gulf Coast Study, criticality evaluation criteria were developed for the following categories:

- **Socioeconomic importance**, such as providing community access to employment centers or mobility for sensitive populations;

- **Use and operational characteristics**, such as vehicle miles traveled and tonnages of freight throughput;

- **Health and safety role in the community**, such as serving as evacuation routes, connecting to important health facilities, or being part of the national defense system.

For each mode, a series of evaluation criteria were developed under each of these categories. Each transportation asset was then evaluated using a combination of qualitative and quantitative factors. Figure 5 shows the highway evaluation criteria used for this project and example scores for a select group of assets. In this case, each asset was scored on a scale of 1 to 3 (3 being most critical), and then the scores were averaged across the criteria. A final designation of criticality was made by ranking each asset by score and then assigning High, Medium, and Low criticality designations by dividing the ranked assets into three equal bins. However, this approach represents just one way to score assets across the criteria. Other practitioners may wish to calculate weighted averages, or set scoring thresholds above which all assets are considered highly critical.

Figure 5: Example Criticality Scoring For Selected Highway Assets

Facility	Socioeconomic - Locally Identified Priority Corridors	Socioeconomic - Functions as Community Connection	Socioeconomic - System Redundancy	Socioeconomic - Serves Regional Economic Centers	Operational - Functional Classification (Interstate, etc.)	Operational - Usage	Operational - Intermodal Connectivity	Health & Safety - Identified Evacuation Route	Health & Safety - Component of Disaster Relief and Recovery Plan	Health & Safety - Component of National Defense System	Health & Safety - Provides Access to Health Facilities	Criticality Score: (L - Low, M - Medium, H - High)
Airport Blvd (West of Snow Rd)	1	1	1	1	2	2	1	3	1	1	1	L
Airport Blvd (East of Snow Rd)	1	3	1	3	3	3	3	3	1	1	2	H
Argyle Rd	1	1	2	1	1	1	1	1	1	1	1	L
Beauregard Street	1	1	1	2	3	2	3	3	3	1	1	M
Bel Air Blvd	1	1	2	1	2	2	1	1	1	1	1	L
Bellcase Rd	1	1	2	1	2	2	1	1	1	1	1	L
Bellingrath Rd (South of Industrial Rd)	1	1	2	3	2	2	1	3	1	1	1	M
Bellingrath Rd (North of Industrial Rd)	1	1	2	1	2	2	1	3	1	1	1	L
Beverly Rd	1	1	2	1	1	1	1	1	1	1	1	L
Broad Street (North of Spring Hill Ave)	1	1	1	3	3	2	2	3	3	1	1	M

The results of applying this methodology to Mobile are shown in the text box below.

Summary of Results in Mobile: Criticality

The criticality assessment determined which highways, ports, airports, rail, transit, and pipeline facilities were highly critical to Mobile. A summary of these findings is detailed in the table below.

Mode		Critical Assets	Total Assets
Highways		152 miles ■ 71 bridges/culverts	644 miles ■ 630 bridges/culverts
Ports		23 ports	61 ports
Airports		2 airports	17 airports
Railroads		5 facilities/lines ■ 347 miles	14 facilities/lines ■ 590 miles
Transit		2 facilities ■ 1 fleet	2 facilities ■ 1 fleet
Pipelines		426 miles	652 miles

5.1.2. Lessons Learned on Methodology

There were a number of lessons learned in the Gulf Coast Study that may help inform similar criticality assessments performed elsewhere, including:

■ **First, adhering too rigidly to a scoring system could leave out areas of local or cultural importance** that might not otherwise score highly against the other criteria. There may be locations that are essential to a community's identity or provide a difficult-to-quantify benefit to the community. In Mobile, the city of Bayou la Batre provided one such example. Because Bayou la Batre scored low in measurements of population, economic activity, operational usage, and other factors, its transportation assets did not appear to be highly critical. However, in meetings with local stakeholders, the project team learned the importance of this community to the local fishing industry and Mobile's identity; the criticality of assets in this area was subsequently revisited. This example also highlights the **importance of vetting the results of the quantitative analysis with a variety of stakeholders** to ensure that essential assets are being captured.

■ In some cases, **it may be appropriate to identify characteristics that automatically confer a high criticality score**. In the Gulf Coast Study, all criteria were weighted equally, including the role of assets in emergency evacuation plans. As a result, some local emergency evacuation routes were not considered to be highly critical because they did not score highly under the other criteria (e.g., average daily volume). The fact that evacuation routes could be considered to have low or medium criticality seemed counterintuitive to stakeholders. To address this concern, future analyses could consider characteristics such as role in emergency evacuation plans to be automatic qualifiers for a designation of highly critical.

▪ Finally, **it is important to remember that the criticality assessment is an assessment of criticality, not vulnerability**. If an asset is considered to not be highly critical, it will not be evaluated for vulnerability, but this does not mean that it *isn't* vulnerable. When considering the final results of the vulnerability assessment, it is important to remember that there may be other assets that are highly vulnerable but that do not show up in the final results because they were screened out based on lower criticality. In the case of Mobile, some local/county roads that in fact may be quite vulnerable did not make it into the final list of critical assets and as a result were not analyzed for vulnerability.

5.1.3. Resources Available for Conducting Criticality Assessments

The following resources may be useful to transportation practitioners conducting a criticality assessment:

▪ The criticality evaluation criteria for each transportation mode used in the Gulf Coast Study. A complete list of these criteria is available in the report titled *Task 1: Assessing Criticality in Mobile, AL.*[8]

▪ The FHWA guidance *Assessing Criticality in Transportation Adaptation Planning,* discussed in Section 4.2, which steps transportation practitioners through the process of evaluating criticality of their assets.

---

[8] This report is available at available at
http://www.fhwa.dot.gov/environment/climate_change/adaptation/ongoing_and_current_research/gulf_coast_study/.

5.2. Gather and Process Climate Information

5.2.1. Overview of Approach

Transportation assets are already exposed to climate stressors, and their vulnerabilities to the current climate are generally understood. What is more challenging to evaluate, however, is how these transportation assets, designed for today's climate, might fare under future conditions. To understand how the critical assets might be vulnerable to climate change, it is first important to understand how the climate may in fact change in the future.

There are a number of climate stressors relevant to transportation that could be affected by climate change, with the primary ones being temperature, precipitation and changes in streamflow, sea level rise, and storm surge and wind conditions from intense storms. Transportation agencies may choose to focus vulnerability assessments on only one or two of these climate stressors, or on all of them.

For the Gulf Coast study, projected changes for all of these climate stressors were estimated, using the following methods:

- Local temperature and precipitation projections were developed using global climate models, with a key input being different scenarios of greenhouse gas emissions. The model outputs were statistically downscaled to represent local conditions.

- Inundation from sea level rise was modeled spatially by considering feasible levels of global sea level rise, adjusting for local uplift and subsidence of land, and determining which areas of land were of low enough elevation to be inundated under those conditions.

- The extent of surge, and associated wind speeds, from various feasible storm scenarios were modeled using the ADvanced CIRCulation (ADCIRC) model and STeady State spectral WAVE model.

These approaches, and the scenarios assumed, are summarized in Table 1.

The reason that different approaches were used to develop each climate stressor relates to the underlying assumptions and capabilities of climate models. Climate models project temperature and precipitation, amongst other variables, based on assumptions of the level of human- and natural-driven greenhouse gas emissions and the resulting concentrations. However, global climate models do not include all processes that contribute to sea level rise, and they do not model future storms. Therefore, other methods were used to project these variables. To capture the breadth of sea level rise projections supported by current science, a review of scientific literature can provide reasonable ranges of global sea level rise, which then can be adjusted to account for local land subsidence and uplift. Similarly, the scientific literature provides insights into how the frequency and intensity of severe storms may change. Therefore, plausible storm scenarios can be developed to represent intensified versions of storms already experienced today, storms similar to those experienced today but with sea level rise added as a variable, or storms that represent both an intensification and increase in sea level. These storm scenarios are then fed into storm surge models to estimate the resulting storm surge, wave heights, and maximum winds.

Table 1: Summary of Climate Information Developed

Climate Stressor	Scenarios	Timeframes	Approach
Temperature	B1, A2, and A1Fi emission scenarios	2010-2039 (near-term) 2040-2069 (mid-term) 2070-2099 (end-of-century)	Projections were statistically downscaled from a variety of global climate model outputs, and compared to the current baseline to estimate change. Projections were developed for numerous variables. Results focused on extremes, such as number of days above 95 degrees instead of average seasonal temperature.
Precipitation & Runoff	B1, A2, and A1Fi emission scenarios	2010-2039 (near-term) 2040-2069 (mid-term) 2070-2099 (end-of-century)	Precipitation projections were calculated using the same approach for temperature.
Sea Level Rise	30 cm (1 ft) of global sea level rise by 2050, and 75 cm (2.5 ft) and 200 cm (6.6 ft) of global sea level rise by 2100		Global sea level rise values were adjusted based on local data on subsidence and uplift of land.
Storm Surge and Wind	11 storm scenarios based on historical storms modeled with different trajectories, intensities, and sea levels	Not applicable	11 storm scenarios were developed using Hurricane Georges and Hurricane Katrina as base storms, and then adjusting certain characteristics of the storms to simulate what could happen under alternate conditions. Storm surge was modeled for each of these storm scenarios using the ADvanced CIRculation model (ADCIRC). ADCIRC also provided estimates of wind speeds. Wave characteristics were simulated using the STeady State spectral WAVE (STWAVE) model.

The results of applying this methodology to Mobile are shown in the text box below.

Summary of Results in Mobile: Climate Projections

The climate information developed for Mobile in this study revealed that temperatures are projected to increase over time, across all variables, regardless of emissions scenario. This results in more extremely hot days – an additional 13 to 36 days annually above 95°F by mid-century (Figure 6).

Figure 6: Projected Number of Days per Year above 95°F in Mobile, AL

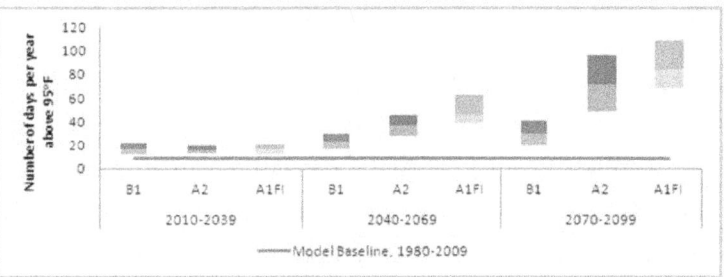

Precipitation projections are less clear-cut. Some variables, such as average annual precipitation, are projected to decrease under some scenarios but increase under others. Under all three scenarios, however, heavy rain events (i.e., the 100-year storm) will become more intense – while the baseline 100-year storm is 12 inches of rain in 24-hours, the 100-year storm may bring up to 18 inches of rain in 24-hours by the end of the century.

The study also found that sea level rise could inundate several areas in Mobile County, with global sea level rise amplified by modest land subsidence over most of the county. The lowest sea level rise scenario of 1 foot (0.3 meters) would inundate about four percent of critical roadway miles and also worsen storm surge from incoming storms. Under a scenario of 6.6 feet (2 meters) of sea level rise, coastal inundation would significantly shift the southern Mobile County shoreline northward, inundate most of Dauphin Island, and flood parts of downtown and the port waterfront. This would inundate 13 percent of critical roadway miles.

Figure 7: Modeled Storm Surge Depth if Hurricane Katrina Had Directly Hit Mobile

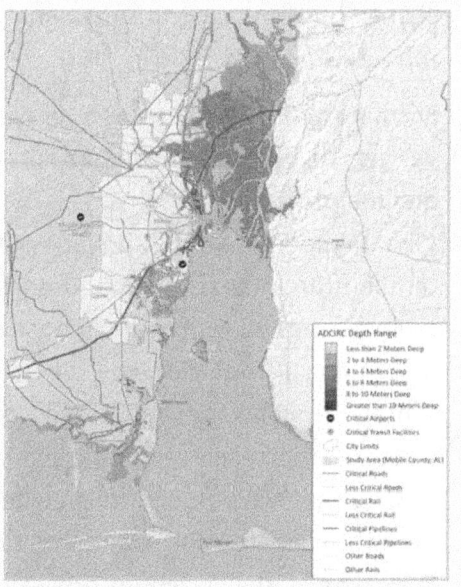

Finally, the study found that transportation infrastructure in Mobile is most exposed to flooding from hurricanes and other tropical storms. If a storm like Hurricane Katrina were to directly hit Mobile, nearly half of the critical roadway miles would be inundated, along with 72% of critical rail links, 92% of critical ports, and 65% of Mobile Downtown Airport (Figure 7). The surge impacts would be worse with any amount of sea level rise. If a more intense coastal storm were to hit Mobile, modeling indicates that the vast majority of critical transportation infrastructure studied could be inundated by depths as high as 38 feet.

5.2.2. Lessons Learned on Methodology

Collecting, interpreting, and utilizing information about projected changes in climate is complicated at best, daunting at worst. However, careful scoping can dramatically reduce the resources needed to develop the information. Transportation practitioners engaging in climate and extreme weather vulnerability assessments may be interested in leveraging the following lessons learned to facilitate their efforts in this area:

- Prior to commencing development of climate data, **it is essential to consider what data types (e.g. timeframes, type of climate stressor) will be most useful for the vulnerability assessment.** Determining the necessary climate inputs prior to conducting the vulnerability assessment can ensure that the needed data is available and can also prevent spending resources developing information that is not needed. For instance, if the assessment is supporting decision making for the short or medium term, then development of long-term climate information (where the choice of scenarios and models becomes more contentious and impactful) may be unnecessary. If the scientific literature indicates that a certain variable may not change considerably in the future, then projections for that climate stressor do not need to be developed. The type of data needed can vary by mode, asset type, and potential impact being evaluated. Literature reviews can help agencies focus on variables that are believed to be changing, and also help eliminate variables that may not change in the future and/or may not be problematic to the local transportation system in the future.

- **Formats of data projections—for example, conveying precipitation projections in daily versus seasonal terms—should correspond to inputs to transportation design, operations, and maintenance decisions.** This is especially true for temperature and precipitation. Climate models routinely provide outputs such as changes in average annual or seasonal temperature or precipitation. Unfortunately, transportation practitioners are generally more concerned with short, intense events, such as heat waves or heavy precipitation that falls during a storm. For the Gulf Coast study, the project team defined a "wish list" of climate variables relevant to transportation decision making. Through discussions with transportation engineers, local and federal transportation planners, operations staff, and climate scientists, along with an extensive literature review, the project team identified desirable climate information that could be derived from daily-scale climate model outputs. For example, the maximum 7-day air temperature is a variable closely tied to pavement design that could be calculated from the data, but that was not a normal climate model output.[9] Similarly, most asset designs are based on estimates of peak streamflow over a

---

[9] Since the climate information was developed for the Gulf Coast study, locally-downscaled data for precipitation and temperature have become available for download online. Furthermore, a tool developed under the Gulf Coast study processes the raw data into more user-friendly formats that resonate with transportation practitioners. Please see discussion of the CMIP Climate Data Processing Tool on page 18.

short period, suggesting that resources should be dedicated to developing projections for short intense events, and when possible, modeling the associated peak flows in a watershed model, rather than focusing on estimating changes in seasonal precipitation.

- In a related point, **it is important to engage transportation engineers when determining the climate data to develop.** The engineers who specialize in certain aspects of design (e.g., hydraulic, structural, and coastal engineers) will have insights into the types and formats of input data that are needed to evaluate potential vulnerabilities of existing or planned assets to future climate stressors. Furthermore, involving engineers at this stage of the project fosters understanding and mutual commitment at later stages; all parties are aware of the limitations and uncertainties associated with the data, and are empowered to use the information in their analyses.

- **"More" is not necessarily "better" when it comes to climate projection data.** A commonly-cited barrier to conducting vulnerability assessments is lack of detailed climate data, but it is also possible to develop too much data. The level of detail and number of scenarios developed in the Gulf Coast Study not only took calendar time and resources to develop, but the volume of information was difficult to analyze and synthesize into "big picture" findings useful for decision makers. For practicality, when it came to screening for and ultimately investigating vulnerabilities of various assets, only a relatively small number of data points and sea level rise and storm scenarios were ultimately used. The combination of thresholds of sensitivity or failure and range of climate projections provided a strong indication of potential vulnerabilities. Zeroing in on a limited number of timeframes, emission scenarios, and storm and sea level rise scenarios can be beneficial in terms of developing the big-picture take-away. Ideally, a small number of scenarios can be used to bracket the range of plausible futures, instead of attempting to consider many different interim scenarios.[10]

5.2.3. Resources Available for Developing Climate Information

Several new resources have become available since climate information for the Gulf Coast study was developed. Statistically downscaled climate data for the entire country are now available, and sea level rise scenarios have been modeled for most of the United States coastline. These resources will dramatically reduce the time and cost involved in obtaining projected climate information at the local level. This section highlights a few key resources now available.

These resources allow transportation practitioners to obtain projected temperature and precipitation data within a few hours rather than the months needed for the Gulf Coast Study.

[10] In situations where transportation agencies are considering *implementing* costly adaptation measures or measures that will commit the agency to a particular course for a long time (e.g., 50 years), it may be necessary to revisit the climate data produced for the purpose of vulnerability screening and develop or expand the suite of scenarios to ensure that investments are robust across a wider range of climate projections.

Downscaled Climate Data Resources for Temperature and Precipitation

There are two important new resources for temperature and precipitation projections that take advantage of the downscaled data now available: the U.S. DOT CMIP Climate Data Processing Tool and the USGS National Climate Change Viewer.

The **CMIP Climate Data Processing Tool**, developed under this study and discussed in Section 4.2, calculates detailed temperature and precipitation variables from raw climate model data. The outputs are designed to present climate projection data in formats that are relevant to transportation practitioners. Example outputs of the tool are shown in the text box below.

The U.S. DOT CMIP Climate Data Processing Tool is available on the FHWA Virtual Framework.[11]

Example Temperature and Precipitation Outputs of CMIP Climate Data Processing Tool

TEMPERATURE

Annual Averages

- Average annual daily mean, minimum, and maximum temperature

Extreme Heat

- Hottest temperature of the year
- 95th and 99th percentile temperature
- Number of days per year and season above 95, 100, 105, and 110 degrees
- Maximum number of consecutive days per year above 95, 100, 105, and 110 degrees
- Highest 4- and 7-day average temperatures

Extreme Cold

- Coldest temperature of the year 1st and 5th percentile temperature
- Number of days per year below freezing
- Lowest 4- and 7-day average temperatures
- Average number of times minimum temperatures fluctuate around freezing

PRECIPITATION

- Total annual precipitation
- 95th and 99th percentile 24-hour precipitation
- Seasonal precipitation
- Largest 3-day precipitation event per season
- Annual maximum 24-hour precipitation (time series)

---

[11] http://www.fhwa.dot.gov/environment/climate_change/adaptation/adaptation_framework/modules/index.cfm?moduleid=4.

The U.S. Geological Survey (USGS) also recently released a web viewer of climate change projections at the county and state level, based on statistically downscaled CMIP5 data. The "**National Climate Change Viewer**"[12] includes projections of high temperatures, low temperatures, precipitation, runoff, snow, soil water storage, and evaporative deficit. For each variable, it provides annual and monthly averages and the 10th, 25th, 50th, 75th, and 90th percentiles for near-term, mid-term, and end-of-century.

Figure 8: USGS National Climate Change Viewer

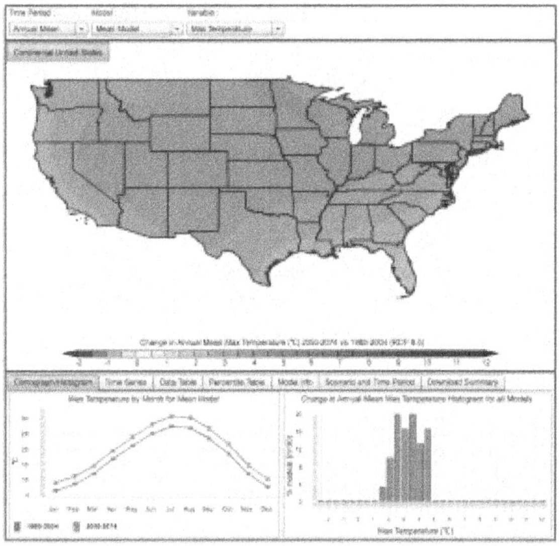

The table below summarizes the U.S. DOT CMIP Climate Data Processing Tool and the USGS National Climate Change Viewer. Each tool, or a combination of both, may be appropriate in different circumstances. For example, the CMIP Climate Data Processing Tool may be more appropriate for studies focused on changes at the local level and on specific transportation-oriented temperature and precipitation variables. The USGS National Climate Change Viewer may be more appropriate for larger geographic areas, studies interested in near-term projections, and studies focused on water-balance variables.

Table 2: Summary of New Downscaled Climate Data Resources

	U.S. DOT CMIP Climate Data Processing Tool	**USGS National Climate Change Viewer**
Underlying data	Bureau of Reclamation Downscaled CMIP3 and CMIP5 Climate and Hydrology Projections	NASA NEX-DCP30
CMIP phase	CMIP3 and CMIP5	CMIP5
CMIP data spatial resolution	1/8 degree (~7.5 mi x 7.5 mi)	800 m (0.5 mi)
Output resolution	Local level – from 56 to 225 sq. miles (between one and four 1/8-degree grids)	County or State level
Output variables	58 temperature and precipitation variables (see box above), including 95th and 99th percentile temperature and precipitation	Minimum temperature, maximum temperature, precipitation, runoff, snow, soil water storage, and evaporative deficit. For each, monthly and annual averages and 10th, 25th, 50th, 75th, and 90th percentiles

---

[12] Available at http://www.usgs.gov/climate_landuse/clu_rd/nex-dcp30.asp

	U.S. DOT CMIP Climate Data Processing Tool		USGS National Climate Change Viewer
Time periods	**CMIP5:** User-defined baseline and future time periods (covering 1950-2099) **CMIP3:** *Baseline:* 1961-2000 • *Mid-century:* 2046-2065 • *End-of-century:* 2081-2099		*Baseline:* 1980-2004 • *Near-term:* 2025-2049 *Mid-century:* 2050-2074 • *End-of-century:* 2075-2099
Models	**CMIP5:** 21 models **CMIP3:** 9 models		**CMIP5:** 33 models
Emissions scenarios*	**CMIP5:** RCP 2.6; RCP 4.5; RCP 6.0; RCP 8.5	**CMIP3:** B1; A1B; A2	**CMIP5:** RCP 4.5; RCP 8.5

* These emissions scenarios come from the IPCC Fifth Assessment Report and Special Report on Emissions Scenarios for CMIP5 and CMIP3, respectively. Additional information on these scenarios is available from the IPCC. Scenarios are listed in order of increasing greenhouse gas concentrations.

Sea Level Rise Resources

Several resources are available for communities looking to model the effects of sea level rise. In addition to localized modeling that may already be available for specific locations, communities can use geospatial bathtub models like the one used in the Gulf Coast study or the **Sea Level Affecting Marshes Model (SLAMM)**. For communities that do not want to undertake their own modeling effort, the National Oceanic and Atmospheric Administration (NOAA) Coastal Services Center released the **Sea Level Rise and Coastal Flooding Impacts Viewer**[13] (see Figure 9). The viewer shows depth and extent of inundation for the entire United States coastline for six sea level rise scenarios, ranging from 1 foot to 6 feet (in 1-foot increments). Users can download the data in GIS form for use in vulnerability assessments or other applications. Also available is the U.S. Army Corps of Engineers (USACE)'s **Sea-Level Change Curve Calculator**,[14] which quickly estimates the relative sea level rise at a given location for each year until 2100, assuming certain rates of sea level rise and local land subsidence/uplift.

---

[13] Available at http://www.csc.noaa.gov/digitalcoast/tools/slrviewer
[14] Available at https://corpsclimate.us/ccaceslcurves.cfm

Figure 9: NOAA Sea Level Rise and Coastal Flooding Impacts Viewer

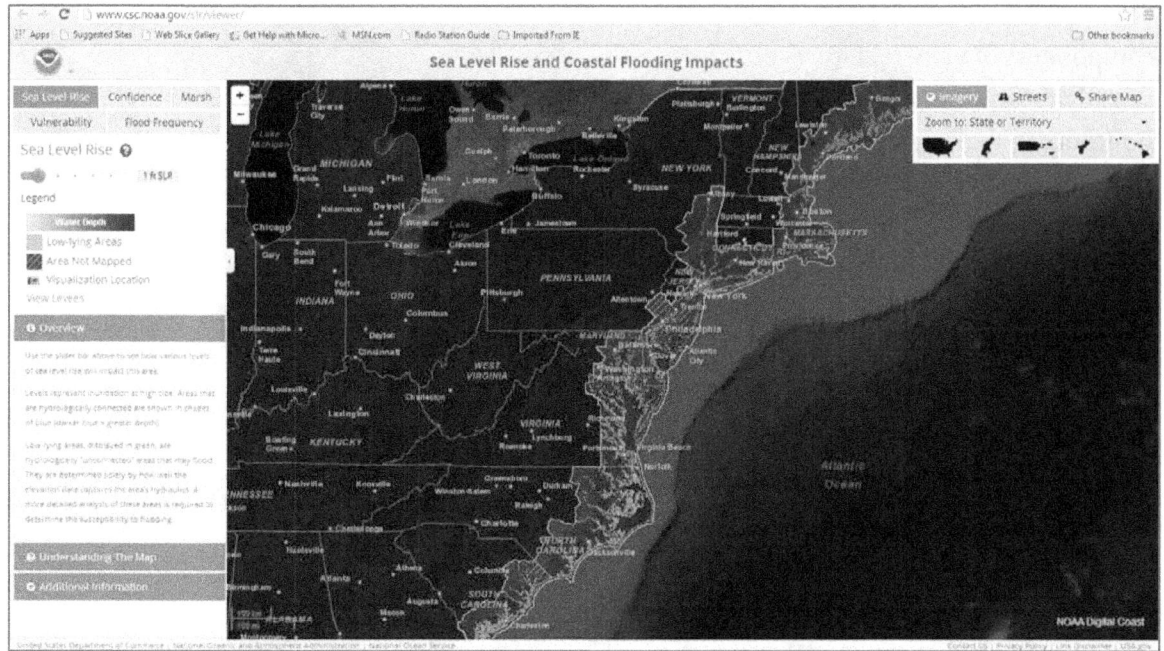

5.3. Assess Vulnerability

5.3.1. Overview of Approach

Detailed vulnerability assessments can be conducted on specific assets. However, when looking at an entire transportation system, comprised of hundreds or thousands of assets, it is often not feasible to conduct detailed assessments of all critical assets. Therefore, it is valuable to conduct a high-level vulnerability screen and assessment to determine (a) system-level vulnerabilities (e.g. which modes, geographic regions, time frames, climate stressors may be of concern) and (b) which critical assets may be more likely to be vulnerable, and thus worthy of a more detailed look.

In order to conduct this type of vulnerability assessment, it is useful to look for relevant characteristics of the assets, surrounding areas, and projected changes in climate that may indicate potential vulnerability. Looking at these "indicators" can allow for a relatively efficient way of evaluating vulnerability across a large number of assets.

In the Gulf Coast study, indicators were developed for each of the three components of vulnerability: exposure, sensitivity, and adaptive capacity. Appropriate indicators were identified through research and consultation with modal experts and local stakeholders. For each asset, the indicators were evaluated using readily-available data sets, spatial analysis, stakeholder input, and expert judgment. For example, information on scour condition from the National Bridge Inventory was used to estimate potential sensitivity of bridges to storm surge; as another example, the degree to which a port is reliant on electricity from the grid was used to estimate potential sensitivity to high winds (which can cause power

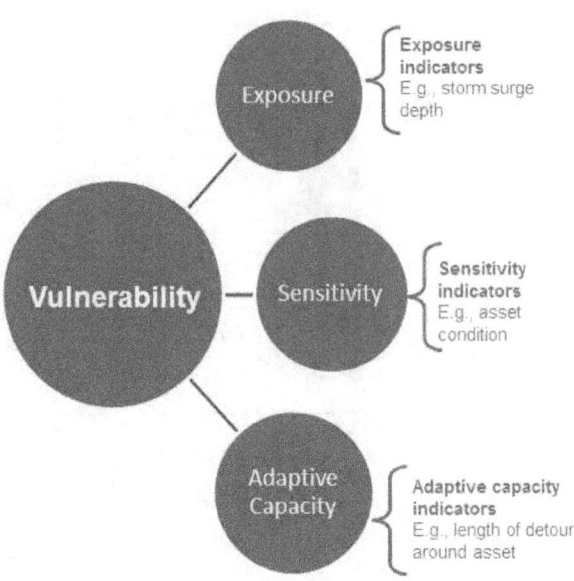

Figure 10: Using Indicators to Assess the Three Components of Vulnerability

outages). Every asset received a numerical score for each indicator, and then a composite score was developed for each asset to estimate their vulnerabilities for each climate stressor, on a scale of 1 (low) to 4 (high).

This approach involved three primary steps:

- **Identify Indicators** – The project team identified exposure indicators for each climate stressor, sensitivity indicators for each climate stressor and transportation mode, and adaptive capacity indicators for each transportation mode. These indicators were selected based on stakeholder input, expert judgment, and data availability. Useful indicators can help distinguish between assets, are based on relatively complete and consistent datasets (across assets being evaluated), and are easily understood and interpreted. In total, this study used 171 indicators to evaluate the vulnerability of Mobile's critical transportation assets.

- **Collect Data on Indicators** – The climate information developed in earlier stages of the project served as exposure indicators for the vulnerability assessment. The project team relied on a combination of nationally available datasets (e.g., the National Bridge Inventory and HAZUS) and local sources to collect information on indicators. In many cases, there were no existing datasets about the indicators, so the project team conducted interviews and surveys with local experts to collect the information.

- **Establish a Scoring Approach** – Finally, the project team developed an approach to convert data on indicators into a single vulnerability score for each asset and climate stressor, in order to review relative vulnerability rankings. This approach used a scale of 1-4 (where 4 is most vulnerable) to rate each indicator, and then calculated a weighted average of indicator scores to determine the vulnerability scores. Stakeholder involvement was key to this approach. The project team developed preliminary scoring bins and indicator weights to reflect relative vulnerability, and then presented the preliminary scoring approach and resulting ratings to local stakeholders to refine. Figure 11 provides an example of how this approach worked to rate the vulnerability of the Cochrane-Africatown USA Bridge to temperature change, using one exposure indicator, three sensitivity indicators, and three adaptive capacity indicators.

- **Refine and Finalize Indicators and Scoring Approach** – Stakeholders reviewed the initial draft results to make sure they reflected on-the-ground conditions. In a few instances, some results were initially inconsistent with the expectations and experience of local stakeholders and other reviewers, prompting a review of the scoring results for these assets. In some cases, it was determined that an adjustment to the indicators or scoring approach was warranted. For example, the precipitation assessment for highways initially overemphasized location in flood zones. In some cases, assets were located in *coastal* flood zones—where flooding might occur due to high tide or surges, rather than precipitation events. Furthermore, assets were initially considered to be "in" a flood zone if any portion of the asset was in the zone—even if it was a very small portion of the asset. Therefore, some highway assets were initially scored as highly vulnerable that did not, in the opinion of local stakeholders, have characteristics that would make them particularly vulnerable to precipitation. As a result, the methodology was revised so that coastal flood zones were removed from the evaluation of precipitation vulnerability, and flood zone scores were based on percentage of the asset within the zone.

In other cases, however, the review of the scoring approach determined that the results provided were in fact legitimate. For example, the Brookley Field airport is located right next to Mobile Bay, and some reviewers expected it to be vulnerable to sea level rise or storm surge. However, the elevation of the airfield is great enough that the airport would not be inundated under any of the sea level rise narratives examined, and all but the most extreme storm narratives.

Figure 11: Example of Vulnerability Score Calculations:
Vulnerability of the Cochrane-Africatown USA Bridge to High Temperatures

Numbers in parentheses represent indicator or component scores,
percentages represents the indicator or component weights

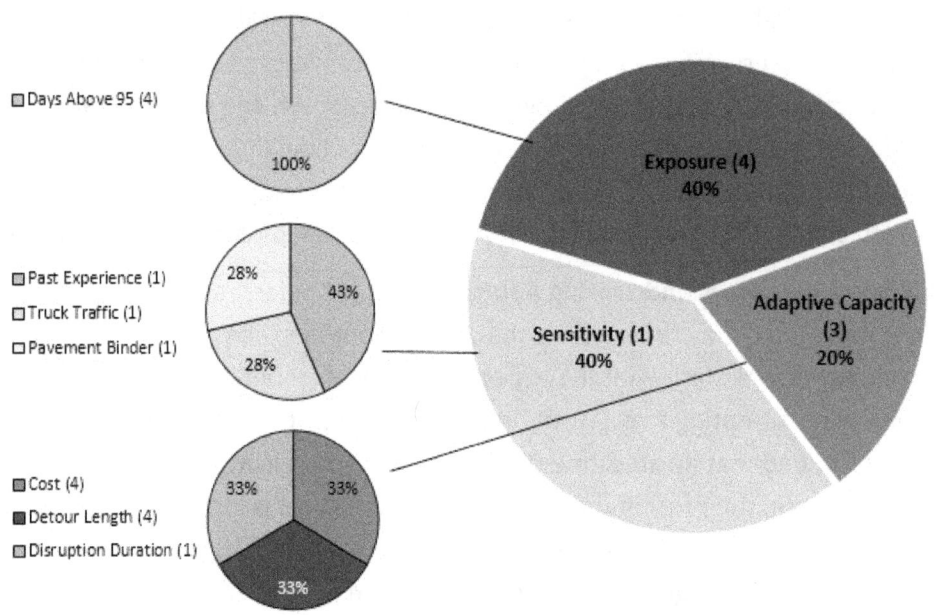

Component	Indicator	Indicator Value	Indicator Score	Indicator Weight	Component Score	Component Weight	Overall Vulnerability Score
Exposure	Days above 95	105	4	100%	4	40%	
Sensitivity	Past Experience	No	1	43%	1	40%	
	Truck Traffic	2723	1	28%			2.6
	Pavement Binder	PG 67-22	1	28%			
Adaptive Capacity	Cost	$210,276,000	4	33%	3	20%	
	Detour Length	65 miles	4	33%			
	Disruption Duration	hours	1	33%			

The indicators approach provides a relatively low-cost way to screen transportation assets for vulnerability by relying on readily available data. The results of the data-driven vulnerability screen provide transportation managers with a starting point for understanding their system's vulnerabilities and making decisions on how to best manage those vulnerabilities. This scoring system allowed for the identification of certain areas that appear particularly vulnerable to certain climate stressors, and for the evaluation of which climate stressors appear to be particularly problematic for certain modes. In addition, the scores allowed for a relative ranking of the assets to determine which assets appear to be the most vulnerable and which ones are less vulnerable.

The indicators approach developed under the Gulf Coast study can be used as a starting point for other transportation practitioners conducting similar assessments. This approach provides flexibility for users to customize the approach with additional/different indicators and/or scoring approaches (e.g., weighting one component or indicator more highly than another) to reflect local priorities and data availability.

The results of applying this methodology to Mobile are shown in the text box below.

Summary of Results in Mobile: Vulnerability Assessment

The vulnerability screen found that sea level rise and storm surge are the most significant climate stressors for the Mobile transportation system. All modes except airports include some assets that are "highly" vulnerable (score ≥ 3) to sea level rise and storm surge. Vulnerability to sea level rise and storm surge is driven largely by exposure – if assets are located within a potential inundation zone, they also tend to be sensitive and thus vulnerable.

Because temperature and precipitation changes were uniform across all assets, those vulnerability rankings reflect differences in asset sensitivity and adaptive capacity.

Several assets emerged as highly vulnerable to multiple climate stressors:

- Alabama State Docks and several other Mobile River ports
- Wallace Tunnel
- SR-193 near the Theodore Industrial Canal
- I-10 bridge across Mobile Bay

In general, highway assets in Mobile were found to be vulnerable across all stressors, while port assets were primarily vulnerable to sea level rise and storm surge. Airports, on the other hand, were most vulnerable to extreme heat, due to the sensitivity of runways and taxiways to heat-related damage. The rail assets evaluated were highly vulnerable to storm surge and sea level rise, but also moderately vulnerable to extreme heat and wind. Finally, the critical transit assets were found to have relatively low vulnerability to all stressors, in large part because the bus system has high adaptive capacity in its ability to reroute around damaged roads, stations, or other challenges.

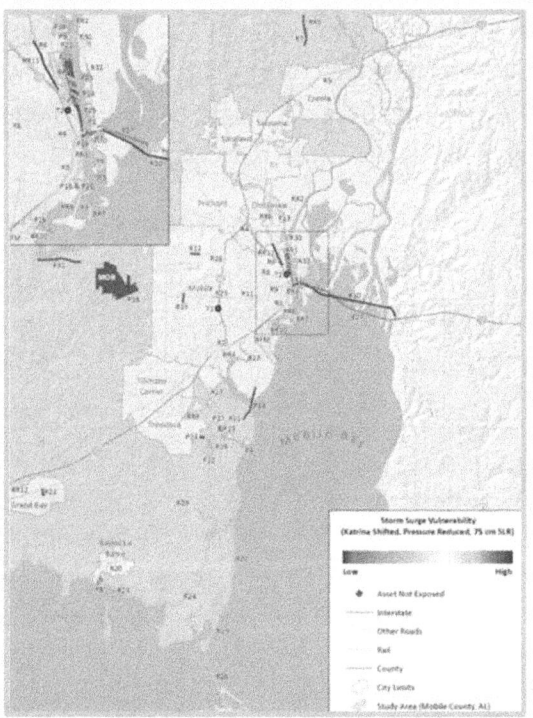

Figure 12: Vulnerabilities of Mobile Transportation Assets to Storm Surge (most extreme scenario)

Geographically, the areas of Mobile County adjacent to the Mobile Bay and to the Gulf of Mexico appear to be particularly vulnerable. This is true not just for storm surge and sea level rise, but also for heavy precipitation. This finding is in line with input from Mobile stakeholders, and with the fact that the coastal areas tend to be lower lying and that some of these areas have existing drainage issues. Some of the assets with particularly high vulnerability to temperature are also near the coast, although their vulnerability is driven by other characteristics—such as their vulnerability to electricity outages or large pavement areas— rather than proximity to the coast. Wind is the only stressor not showing a concentration of vulnerable assets near the coastal regions. The assets with higher wind vulnerability scores extend inland from Downtown, and are also in the more inland, southern part of the County. Some of these segments are in areas with a larger number of traffic signals.

5.3.2. Lessons Learned on Methodology

There were a number of lessons learned in the Gulf Coast Study that may help inform similar vulnerability screenings performed elsewhere, including:

- The indicator-based vulnerability screening approach offers a systematic, transparent approach. However, its limitation is that it will never perfectly capture local circumstances or asset-specific details. Instead, **this approach provides a starting point for understanding relative vulnerability of assets for specific climate stressors, and general vulnerabilities across climate stressors.** From the initial screening results, decision-makers may tweak and/or adjust weighting and selection of indicators to reflect local circumstances. Further analyses can be undertaken to understand case-by-case vulnerabilities for assets of concern.

- **Perfect information is not necessary**, even for an indicator-based vulnerability assessment. Using what data are available for an initial screen can provide a relatively quick and easy way to get to a starting point for understanding vulnerability and engaging stakeholders.

- **Preliminary results should be vetted with knowledgeable, local stakeholders.** Identifying unexpected results can help refine and improve the indicators and scoring systems. Maintenance staff and local engineers are likely to have the most up to date information on vulnerabilities in the system.

- **Determining a single vulnerability score for each asset is convenient and useful, but can hide some nuances of vulnerability.** An asset's vulnerability score covers many factors, such as whether it will experience the stressor, how it will respond to the stressor, and the broader implications of that response. A single score simplifies these factors, and a single score does not communicate what is driving the results. For example, in the Gulf Coast project, the use of asset replacement cost to approximate adaptive capacity skewed some vulnerability results towards more expensive assets, although they were not necessarily more exposed or more sensitive to climate change than less expensive assets. Other ways to view the results can be considered to address this, such as the "Damage" vs. "Adaptive Capacity" approach presented in the Vulnerability Assessment Scoring Tool (see page 30).

- Similarly, transportation practitioners should not put too much stock in minor differences in absolute scores of assets. If a scale of 1-4 is used to express vulnerability, for instance, an asset that scores as a 3.3 is not necessarily more vulnerable than an asset scoring 3.2. There will inevitably be a margin of error in the results. Rather, **it is important to use the results to identify "big picture" findings of system- and community-wide vulnerabilities, and to get a general sense of which assets may be more vulnerable and which ones may not.** In this project, sensitivity analyses were used to consider how certain indicators (e.g., current replacement value) may be appropriately (or inappropriately) driving the ultimate scores and ranking of vulnerability.

- **Data collection can be difficult and time-consuming**, but if done strategically, it can have value for an agency beyond the vulnerability assessment. For example, state DOTs may find it helpful to coordinate between data collection for vulnerability indicators and ongoing asset management system efforts.

- **Asset Management Systems** are good places to include information on current and projected vulnerabilities for specific asset locations. Asset managers, planners and engineers can then readily access the information when it comes time to consider retrofits, rebuilding, etc.

5.3.3. Resources Available for Assessing Vulnerability

The **Vulnerability Assessment Scoring Tool (VAST)**, developed under the Gulf Coast Study and discussed in Section 4.2, is an Excel spreadsheet that assists users with conducting an indicator-based vulnerability screen. It provides the framework for entering assets to be evaluated and selecting relevant climate stressors and indicators, and then calculates vulnerability scores. (see Figure 13 for an example input page of VAST).

The **Sensitivity Matrix**, also discussed in Section 4.2, provides information on how transportation assets might be sensitive to different climate stressors.

Resources available to help evaluate indicators include geospatial analysis, the National Bridge Inventory, data from transportation agency asset management systems, and many others. Please refer to the Gulf Coast study's Task 3.1 report for more information on the data sets used for that vulnerability assessment.

Figure 13: Screenshots from the U.S. DOT Vulnerability Assessment Scoring Tool (VAST)

(a) The VAST "Indicator Library" provides examples of indicators, potential data sources for each indicator, and example ways to convert data on that indicator into scores; (b) VAST allows users to adjust indicator weights; (c) VAST outputs vulnerability scores for each asset and stressor

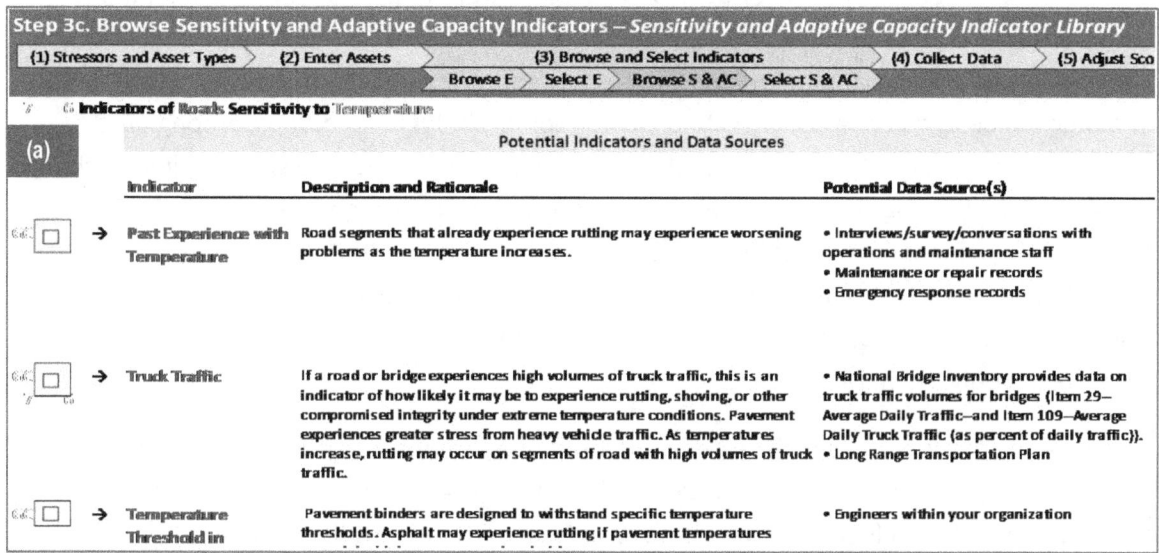

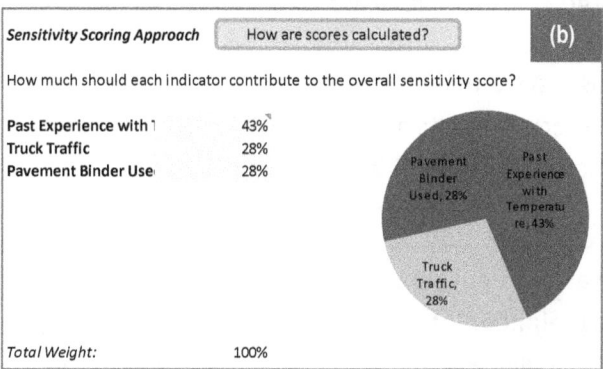

(c)	Temperature Changes			
	Warmer Scenario	Warmer Scenario	Hotter Scenario	Hotter Scenario
Asset ID	"Damage"	Vulnerability	"Damage"	Vulnerability
R27_B9	1.5	1.8	2.5	1.8
R5_B9	1.5	1.6	2.5	1.6
R26_B1	1.5	1.8	2.5	1.8
R30_B1	1.5	1.8	2.5	1.8
R14_B1	1.5	1.7	2.5	1.7
R10_B3	1.5	1.5	2.5	1.5
R10_B4	1.5	1.6	2.5	1.6
R16_B1	1.5	1.6	2.5	1.6
R10_B12	1.5	1.5	2.5	1.5
R10_B2	1.5	1.5	2.5	1.5
R4_B4	1.5	1.5	2.5	1.5

5.4. Conduct Detailed Engineering Assessments

5.4.1. Overview of Approach

Vulnerability screens are useful for developing a big-picture understanding of vulnerability within a transportation system, and to understand which specific assets may be particularly vulnerable to climate change. However, across transportation assets, there is a wide variation in materials, design standards, and site-specific geomorphologic conditions, among other characteristics—all of which influence whether an asset is vulnerable to specific climate change stressors. The full range of these details cannot be captured in a high-level screen. Looking at the specific engineering and surrounding site conditions, on the other hand, can provide a more accurate picture of an asset's vulnerability. It can also allow transportation practitioners to discuss the efficacy of specific adaptation options, rather than talking about adaptation measures in only a general manner.

The resource requirements of detailed engineering assessments make it infeasible to conduct them for a large number of assets. Therefore, engineering assessments might be conducted after a system-level vulnerability assessment and screen that identifies a small set of potentially vulnerable assets. Or, transportation agencies could do asset-level assessments for structures even without conducting a system-level screen first, in situations where certain assets are particularly critical or show signs of potential vulnerability.

However, project-level assessments are not as simple as plugging a new number into traditional engineering calculations, particularly since climate projections usually come with a range of values or inherent uncertainty. In the Gulf Coast study, a flexible 11-step *General Process for Transportation Facility Adaptation Assessments* (the *Process*) was used as a framework for conducting detailed engineering assessments. These steps are shown in the text box to the right.

> **11-Step General Process for Transportation Facility Adaptation Assessments**
>
> 1. Describe the Site Context
> 2. Describe the Existing/Proposed Facility
> 3. Identify Climate Stressors that May Impact Infrastructure Components
> 4. Decide on Climate Scenarios and Determine Magnitude of Changes
> 5. Assess Performance of Existing/Proposed Facility
> 6. Identify Adaptation Option(s)
> 7. Assess Performance of Adaptation Option(s)
> 8. Conduct an Economic Analysis
> 9. Evaluate Additional Decision-Making Considerations
> 10. Select a Course of Action
> 11. Plan and Conduct Ongoing Activities (including monitoring performance of selected adaptation strategy)

This process was applied to 11 case studies, which are described in Table 3. Within each case study, methodologies were developed and employed to evaluate how the projected changes in climate (developed earlier in the project) could affect the structural integrity of the asset. The effectiveness of specific adaptation measures was also evaluated.

Table 3: Detailed Engineering Assessments Conducted Under the Gulf Coast Study, and Example Key Findings

Climate Stressor	Asset Type	Damage Mechanism	Asset Location	Example Key Finding
Increased precipitation	Culvert	Flooding	Airport Boulevard culvert at Montlimar Creek	Benefit-cost analyses of adaptation options are greatly influenced by what is included within the bounds of the analysis. It is up to the analyst to determine which benefits will be included in the analysis; however, the case study does suggest a need to consider benefits beyond the road right-of-way such as damages to businesses and buildings.
Sea level rise	Navigable waterway bridge	Clearance for navigation	Cochrane-Africatown USA Bridge	Given the wide range of feasible adaptation options, port and transportation planners should begin monitoring sea level rise and its potential constraints on ship navigation as soon as possible. It may be decided that immediate action is not needed, but understanding future constraints could be factored into decisions related to siting of port facilities and upcoming bridge rehabilitation processes.
Sea level rise	Bridge approach embankment	Slope erosion	West approach embankment of the US 90/98 Tensaw-Spanish River Bridge	Any protection recommended for a bridge approach embankment like the one studied would need to address all potential stressors upon the abutment including storm surge and scour.
Higher storm surge	Bridge abutment	Abutment scour	West abutment of the US 90/98 Tensaw-Spanish River Bridge	Inspectors should be informed that even if the structural portion of an abutment is situated on "dry" ground, other components such as bulkhead, riprap, or other stability measures may play a key role in the overall scour resistance of the abutment and should be monitored.
Higher storm surge	Bridge segment	Wave forces, bridge pier scour	US 90/98 ramp to I-10 eastbound at Exit 30	The worst case storm surge scenario does not necessarily translate to the worst effect on the facility due to the relative effects of wave impacts damage under different scenarios.
Higher storm surge	Road alignment	Overtopping / slope erosion	I-10 (mileposts 24 and 25)	Additional erosion protection should be considered when designing roadway crossings that could be subjected to reverse flow from storm surges.

Climate Stressor	Asset Type	Damage Mechanism	Asset Location	Example Key Finding
Higher storm surge	Coastal tunnel	Flooding	I-10 (Wallace) Tunnel	When evaluating the impacts of storm surge, wave height must be considered waves transmit localized, high impact forces that can be particularly damaging to structures that aren't normally exposed to (and hence, not designed for resisting) these forces. Additionally, the most commonly understood measure of storm strength - the Saffir-Simpson Hurricane Wind Scale, which assigns wind speeds to certain categories (e.g. a Category 5 hurricane) - is not particularly useful for engineering analyses for storm surges because there is no one-to-one relationship between storm surge and storm "category."
Higher storm surge	Shipping pier	Wave forces	McDuffie Coal Terminal, Dock 1	Loads used to design piers are beyond expected lateral loads expected from even the most extreme storm surges, although the equipment and structures on top of the dock are still vulnerable.
Temperature changes	Roadway pavement	Rutting and cracking	Generic location	It would be better to monitor temperature changes, periodically update historical temperature records, and use climate projections where appropriate rather than existing historical data currently used by pavement design software.
Temperature changes	Continuously welded rail	Buckling, pull-aparts	Generic location	The neutral temperature for railroad track would be inadvisable under the "Hotter" narrative Scenario at all future time periods. Continuing to use the adopted neutral temperature might increase the risk of sun kinks in the future.
Precipitation, wind, temperature, sea level rise, hurricanes	Operations and maintenance activities for various facilities	Varies based on climate stressor	Alabama Department of Transportation, City of Mobile, and Mobile County operations and maintenance practices	O&M personnel in the Gulf Coast region and elsewhere need to be prepared for the unique and continuing challenges of extreme weather particularly when it comes to cooperation between organizations.

5.4.2. Lessons Learned on Methodology

There were a number of lessons learned in the Gulf Coast Study that may help inform similar engineering assessments performed elsewhere, including:

- **The 11-step *Process* can be successfully applied across different types of assets and for a range of climate change stressors.** The *Process* provides a consistent analytical approach across the various engineering disciplines involved in the analyses for this study. It can therefore serve as an organizing framework for how engineering design can be undertaken considering the uncertainties associated with possible future environmental conditions.

- **A design process that reflects projected changes in climatic conditions has to account for possible changes in the input values of the design variables beyond simply relying on historical data.** This is a significant shift from standard engineering design practice. In order to do so, input data must be provided at a scale appropriate to the design process. A lack of useful data has been a challenge noted for many years and is an identified gap in the application of climate scenario projections in engineering design. This study strove to develop data at the temporal and spatial scale needed to conduct engineering design at the project level as highlighted in Section 4.2 and 4.3.

- **For the sake of a robust design process, it is important that a range of climate change-related variables be considered**, simply to make sure that even the lower estimates do not require corrective design action, and that a reference alternative is presented for the scenario analyses of the higher stresses on the assets. Additionally, in some cases, the lower scenario was actually found to be more damaging than the higher scenario. For example, since wave impacts tended to be more damaging than the storm surge itself, some assets actually could experience less damage if they were completely submerged than if they were partly submerged and still subject to wave impacts.

- **Basing future scenarios on the experience of a single historical weather event and then altering characteristics to reflect possible future permutations provides relatable results to local stakeholders.** This is especially the case if a severe storm event occurred recently. However, this approach does not allow for the calculation of a return period, which presents a challenge when comparing future asset performance against a design standard based on return periods (e.g., no overtopping is allowed for storms exceeding the 100-year storm).

- **A Monte Carlo analysis can be an effective way to deal with the climate-related uncertainties influencing major projects.** In the Airport Boulevard Culvert study, a Monte Carlo analysis was used to simulate thousands of different combinations of precipitation events under five climate/land-use scenarios and then estimated the resultant flooding costs over a 30-year appraisal period. This approach was very useful in considering future climatic conditions that could affect the performance of the culvert and therefore of the benefits associated with adaptation options.

5.4.3. Resources Available for Conducting Detailed Engineering Assessments

The case studies developed under the Gulf Coast project contain detailed information on how they were conducted and the results of the analyses with the intention that they could be replicated elsewhere. Each case study followed the same general 11-Step Process with the specific methodologies tailored to each asset-stressor combination. The 11-Step Process can be followed by other transportation practitioners seeking an in-depth understanding of specific assets' vulnerabilities and which adaptation strategies might be effective. The specific methodologies in each case study can also be used as a starting point for practitioners conducting analyzes for similar asset-stressor combinations. The details of the case studies and additional information on the 11-Step Process can be found in *The Gulf Coast Study, Phase 2 Engineering Analysis and Assessment Final Report, Task 3.2*.

6. Areas for Future Work

The current methodologies, indicators, and tools discussed in this report offer important steps towards assessing the risk posed by future changes in climate and identifying adaptation options, but there are still areas that would benefit from additional research and evaluation. These areas for future work, as well as current initiatives that may address some of these gaps, are discussed below.

6.1. Develop methodologies for assessing costs and benefits of adaptation options

In many ways, state DOTs are run like businesses, and all investments must be able to be justified as financially sound. Decisions about which adaptation measures to undertake must be viewed within the larger context of transportation agency budgets, and other priorities of the agency that compete for limited resources. Lack of funding is often cited as a primary barrier to adapting to climate change.

Thoughtful, well-reasoned adaptation strategies can save funding over the long term. Understanding the costs and benefits of adaptation can help show which vulnerabilities should be addressed now and which ones can wait, and it can also help decision makers choose among a range of potential adaptation options. There is limited research on the costs and benefits of adaptation at a local- or project-scale, and also limited guidance on how to assess costs and benefits. More research into this area would be beneficial to transportation practitioners grappling with the challenge of preparing for a changing climate in the face of tightening budgets and other agency priorities.

One of the detailed engineering case studies used a Monte Carlo analysis to compare the cost-benefits of different adaptation options. However, there were some limitations in the analysis used in the Gulf Coast study. For example, it did not investigate how individual adaptation projects should be considered in the context of a larger state or local investment and asset management programs. With limited resources and implementation difficulties, it will become important for planners and engineers to identify and explain co-benefits of adaptation strategies to increase the understanding of the potential benefits of such projects, thereby increasing the likelihood that well-thought out adaptation treatments will get prioritized and funded.

6.2. Refinement of indicators and scoring approaches

System-scale vulnerability assessments such as the one done for the Gulf Coast study rely upon indicators of criticality, exposure, sensitivity, and adaptive capacity. The indicators developed under the Gulf Coast study are an excellent starting point for other vulnerability assessments, but more research could be done to evaluate the effectiveness of the indicators and associated scoring approaches. For example, looking at past climate events and the associated impacts on the transportation system may help determine if the chosen indicators are actually identifying the most vulnerable assets.

There is a need to develop indicators that cover climate stressors relevant to geographic areas outside of Mobile and the Gulf Coast that cover asset types or engineering practices (e.g. stormwater management best management practices) not evaluated under this project, or that rely on data that may be more readily available in the future and/or for other areas. The methodologies developed in this study were shaped by Mobile-specific priorities and do not include cold-region concerns such as permafrost and snow, detailed discussions of public transit, or certain engineering practices. Other indicators may not have been included in this analysis because data was difficult to obtain in Mobile, but those datasets may be more available in other locations. Further research could be done to develop indicators that can be applied more widely across the United States.

6.3. Improve integration of adaptation efforts into other transportation processes and systems

Climate change should not be thought of in a silo. There are a number of existing asset management systems and regulatory processes that should incorporate adaptation considerations. Transportation agencies will ultimately be able to make more informed decisions regarding climate change, and leverage the resources spent on other assets and processes, if climate change considerations are well-integrated throughout the agency.

For example, the formal transportation planning process includes established processes for planning, siting, and designing transportation assets. These processes are the ones that result in actual projects. When adaptation is considered separately from these processes, it is difficult to have adaptation considerations influence the decisions ultimately made about that project.

Meanwhile, there is already considerable investment in efforts and systems related to asset management and risk management. By integrating adaptation into these efforts, a major barrier to adaptation planning—cost—can be reduced, and adaptation can be considered in a streamlined manner, rather than representing one more administrative process to go through. For example, FHWA is encouraging risk assessment as part of asset management planning in response to the requirements of the Moving Ahead for Progress in the 21st Century Act (MAP-21). FHWA is already taking steps to integrate asset inventories, GIS information, and existing climate risks with traditional stressors like asset age and physical condition. Since *current* climate risks are already being considered, it would be logical to integrate *projected future* climate into these efforts, rather than building an entirely new process for evaluating climate change vulnerabilities and adaptation.

6.4. Coordinated Guidance from Federal Agencies

Local transportation stakeholders participating in the Gulf Coast Study noted that they sometimes received potentially conflicting guidance from federal agencies on how to plan for climate change. Transportation agencies may receive different messages about considering climate change. For example, they may be instructed to take into account flood zones determined by the Federal Emergency Management Agency (FEMA), which expressly do not account for climate change. Meanwhile, although FHWA supports various climate change adaptation project types and encourages transportation officials to consider future flood conditions and projected climate information when planning and designing transportation projects,[15] post-disaster funding is often used to repair or replace damaged assets "in-kind", as some facilities need to be rebuilt as quickly as possible to meet transportation needs. These funding decisions, thus, seem in conflict with other federal initiatives that encourage transportation agencies to build their system to be more resilient to extreme weather and climate.

Improved coordination across federal agencies to convey a consistent message regarding how to prepare for climate change would be beneficial. However, achieving this coordination is no small feat, and in the short-term, it may be necessary to simply develop approaches for preparing for climate change that are not in conflict with current guidance of other federal agencies.

6.5. Complementary U.S. DOT Projects Currently Underway

The U.S. DOT is seeking to build on the existing base of knowledge regarding climate change vulnerability and adaptation assessments. These projects aim to further the research already conducted, and to reduce the barriers discussed in this report and identified under other efforts. For example:

▪ The U.S. DOT recently funded three sets of climate resilience pilot projects, two under FHWA and one under the Federal Transit Administration (FTA). The pilot studies were designed to test and refine methodologies for evaluating vulnerabilities and adapting to climate change and extreme weather.[16] These pilot efforts are building on the lessons learned from the Gulf Coast study as well as other similar studies.

▪ FHWA is also funding a project called *Transportation Engineering Approaches to Climate Resilience*. This work, expected to be completed in early 2016, will include up to 20 detailed

---

[15] For more information, please see FHWA's "Eligibility of Activities to Adapt to Climate Change and Extreme Weather Events Under the Federal-Aid and Federal Lands Program," dated September 24, 2012, at http://www.fhwa.dot.gov/federalaid/120924.cfm .
[16] For more information about these pilot programs, please see http://www.fta.dot.gov/12347_14013.html and https://www.fhwa.dot.gov/environment/climate_change/adaptation/ongoing_and_current_research/vulnerability_assessment_pilots/

engineering analyses focused on evaluating asset-level vulnerabilities and adaptation options. This project will produce additional information, guidance, and tools for integrating climate change considerations into engineering design practices. The project will cover a diverse range of transportation assets and climate stressor combinations across the United States and will specifically focus on closing gaps that currently deter the integration of climate change into engineering design. Similar to the case studies developed under the Gulf Coast study, these analyses will build on previous lessons from this and other projects.

■ Another FHWA-funded project is looking at transportation systems in the New York-New Jersey region to evaluate the impacts of Superstorm Sandy and to improve transportation system resiliency to extreme weather and climate change. This project is due to be completed in Summer of 2015.

■ Finally, FHWA is developing technical guidance and methods (HEC-25 Volume 2: Highways in the Coastal Environment: Assessing Extreme Events) to incorporate sea level rise, storm surge, and wave action into coastal design. This guidance is due in the Fall of 2014.

This suite of initiatives will complement similar activities across the United States at the state and local levels. The state of knowledge is rapidly advancing, providing transportation agencies with more and more resources and guidance for preparing for climate change.